BUGS UP C

ANTS
UP CLOSE

RACHAEL MORLOCK

PowerKiDS
press.
New York

Published in 2020 by The Rosen Publishing Group, Inc.
29 East 21st Street, New York, NY 10010

First Edition

Editor: Elizabeth Krajnik
Book Design: Michael Flynn

Photo Credits: Cover, p. 1 Rav Kark/Shutterstock.com; (series background) Karuka/Shutterstock.com; p. 5 teguh santosa kedua/Shutterstock.com; p. 6 Tomatito/Shutterstock.com; pp. 7, 15 Pavel Krasensky/Shutterstock.com; p. 9 kurt_G/Shutterstock.com; p. 11 Andrey Pavlov/Shutterstock.com; p. 13 pan demin/Shutterstock.com; p. 17 vblinov/Shutterstock.com; p. 19 Darkdiamond67/Shutterstock.com; p. 21 Tacio Philip Sansonovski/Shutterstock.com; p. 22 Tsekhmister/Shutterstock.com.

Library of Congress Cataloging-in-Publication Data

Names: Morlock, Rachael, author.
Title: Ants up close / Rachael Morlock.
Description: New York : PowerKids Press, [2020] | Series: Bugs up close! | Includes index.
Identifiers: LCCN 2019016491| ISBN 9781725307742 (paperback) | ISBN 9781725307766 (library bound) | ISBN 9781725307759 (6 pack)
Subjects: LCSH: Ants–Juvenile literature.
Classification: LCC QL568.F7 M667 2020 | DDC 595.79/6–dc23
LC record available at https://lccn.loc.gov/2019016491

Manufactured in the United States of America

CPSIA Compliance Information: Batch #CWPK20. For Further Information contact Rosen Publishing, New York, New York at 1-800-237-9932.

CONTENTS

Getting Antsy

Unless you live in Antarctica, you've probably seen ants in nature. Scientists have discovered more than 12,000 **species** of ants around the world. These ants come in different sizes, shapes, and colors. Scientists think there are between 1 **quadrillion** and 10 quadrillion ants on Earth!

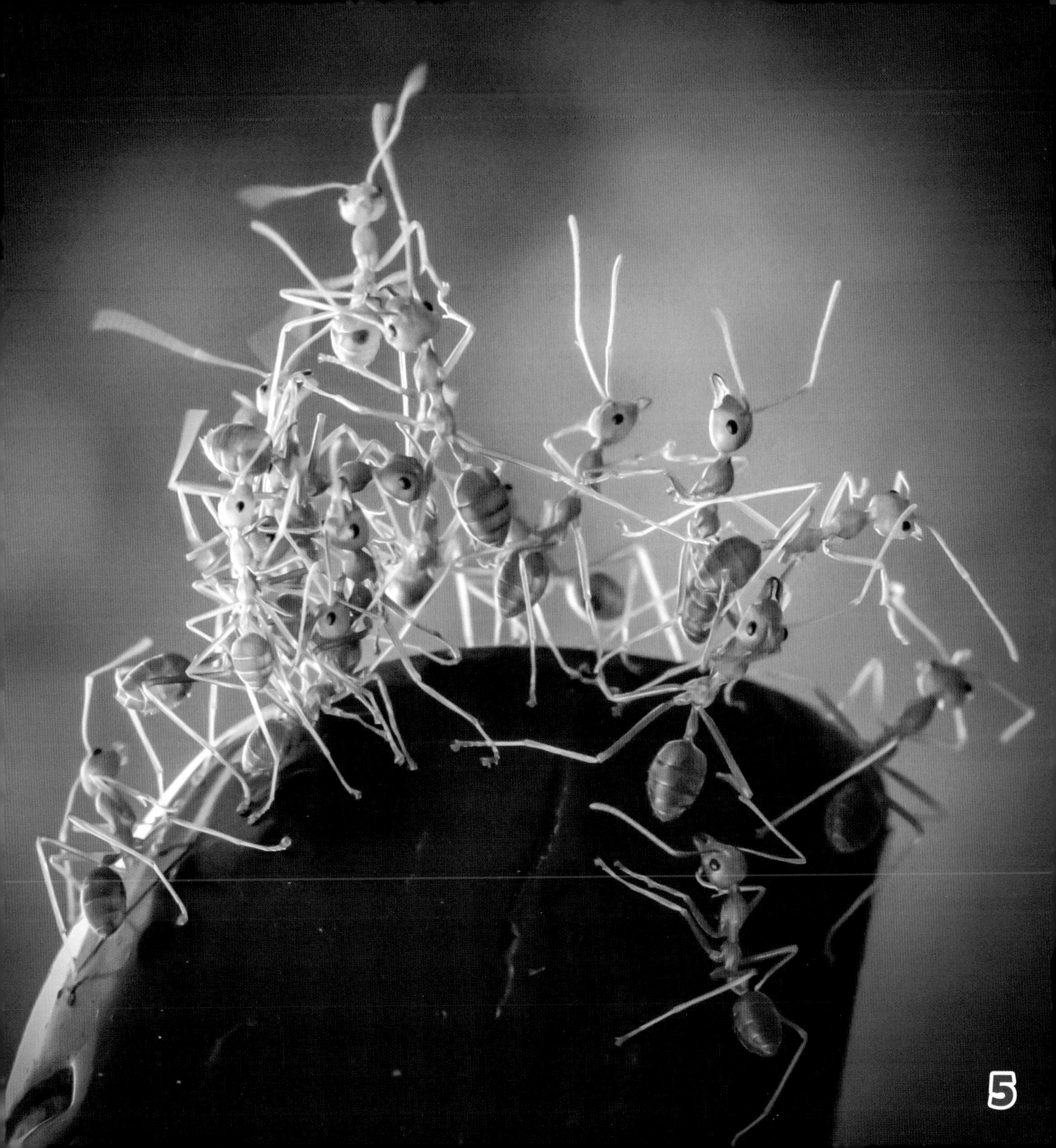

From Egg to Adult

All ants begin life as tiny oval eggs. When an egg hatches, the wormlike creature that comes out is called a larva. The larva grows larger, shedding its skin. Then, the larva spins a **cocoon** around itself. Now, it's a pupa. When the pupa comes out of the cocoon, it's an adult.

Parts of an Ant

An ant's head has antennae, eyes, and mandibles, or mouthparts. Most ants have compound eyes, which means their eyes are made up of hundreds of lenses that make one picture in the **brain**. An ant's antennae help it touch, taste, smell, and move things. Ants use their mandibles to hold things, eat, and fight.

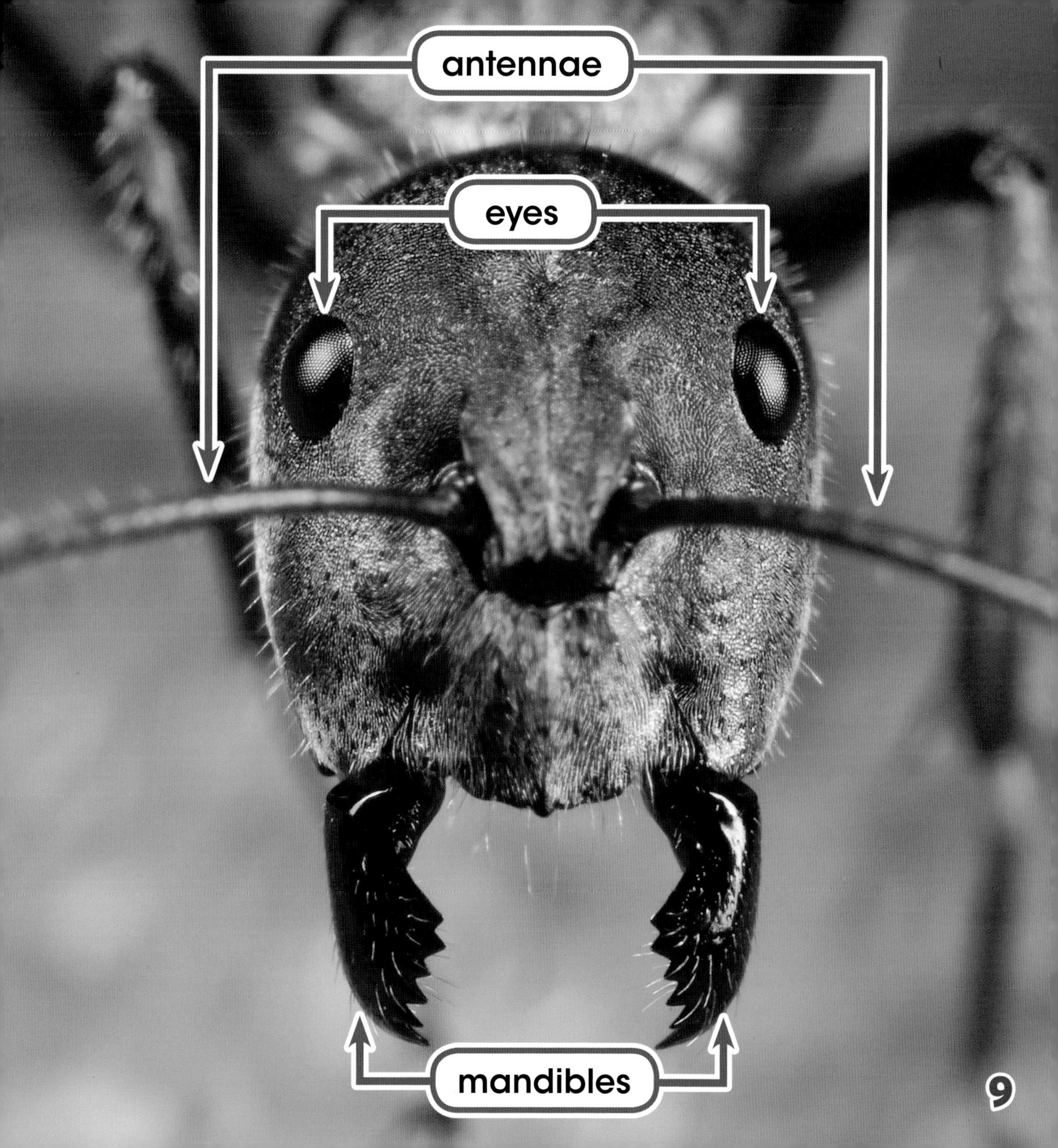
antennae
eyes
mandibles

The ant's middle part is called the thorax. It includes the ant's three pairs of legs. At the end of each leg is a claw. The ant's end part is called the abdomen. This contains the ant's heart and other **organs**. Some ants have a stinger, too.

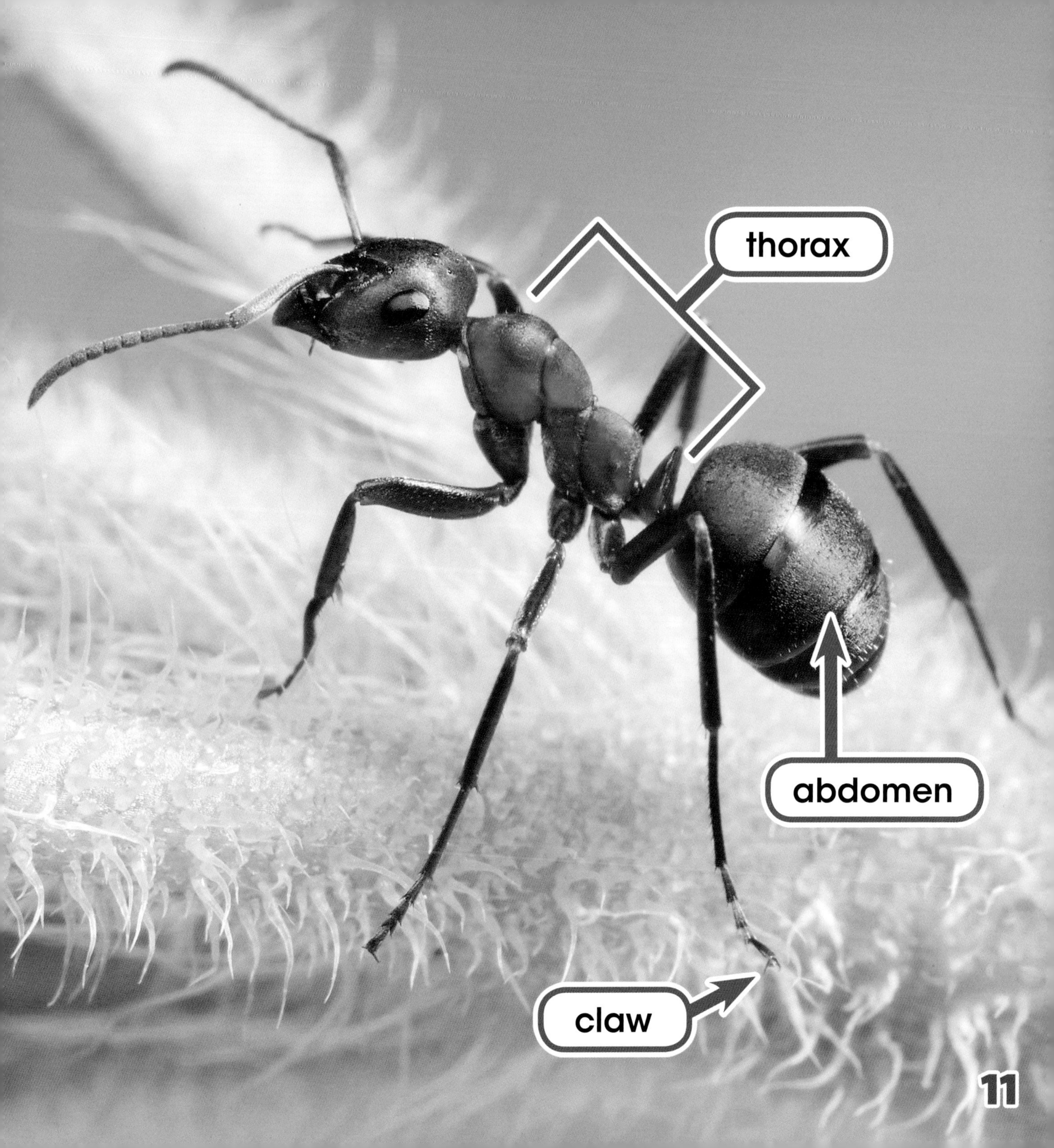
thorax
abdomen
claw

Life in the Colony

Ants live together in groups called colonies. These groups are made up of thousands or even millions of members. The three kinds of ants are queens, drones, and workers. Ants work together to build a nest, find food, care for the queen and her eggs, and **protect** the colony from other ants.

Queen Ants

Queens are the only ants that lay eggs. Most queen ants are born with wings, which they shed after **mating** with a drone. Then the queen finds a good place to start a colony. Queens spend most of their lives in the nest, laying eggs. They live up to 30 years!

queen

Drone Ants

Drones, which are male ants, are also born with wings. They have a large thorax to hold up their wings. Drones are usually smaller than queens, but they have larger eyes and straighter antennae. Drones' only job is to mate with a queen, and they die shortly after mating.

queen
drone

Worker Ants

Worker ants are females that don't lay eggs. They take care of the young, build and clean the nest, gather food, and fight enemies. Worker ants can lift between 10 and 50 times their body weight. They can leave trails of **chemicals** to guide other ants to food sources.

Ants Around the World

Each ant species is different based on where and how it lives. Some sting their enemies to fight them off. Others have strong, sharp mandibles for hunting and gathering food. Scientists think that many ant species haven't even been discovered yet!

Ants in Action

Life on Earth would be very different without ants. Ants carry and spread seeds. They also dig through the earth, which helps plants get the water and air they need to grow. Ants may be small, but they play an important part on Earth.

GLOSSARY

brain: An organ in the head that controls movement, thoughts, feelings, and more.

chemical: Matter that can be mixed with other matter to cause changes.

cocoon: A covering that some insects form around themselves to protect them while they grow into adults.

mate: To come together to make babies.

organ: A body part that does a certain task.

protect: To keep safe.

quadrillion: 1,000,000,000,000,000. (Number 1 followed by 15 zeros.)

species: A group of plants or animals that are all the same kind.

INDEX

WEBSITES

Due to the changing nature of Internet links, PowerKids Press has developed an online list of websites related to the subject of this book. This site is updated regularly. Please use this link to access the list: www.powerkidslinks.com/buc/ants